DEDICATION

This book is dedicated to our loving daughter Jayani.

ACKNOWLEDGEMENTS

We express our gratitude to our parents and in-laws for their constant encouragement, support and blessings.

It will be an injustice if we do not thank all our students for their innovative ideas and feedback.

CONTENTS

Sr. No.	Topic	Page No
1.	General toxicology	5
2.	Agricultural poisons	13
3.	Corrosive poisons	18
4.	Inorganic metallic irritant poisons	22
5.	Inorganic non-metallic irritant poisons	31
6.	Organic plant irritant poisons	33
7.	Organic animal irritant poisons	36
8.	Somniferous poisons	40
9.	Inebriant poisons	43
10.	Deliriant poisons	52
11.	Cardiac poisons	56
12.	Spinal poisons	58
13.	Asphyxiants	60
14.	Food poisons	63
15.	Drug dependence and abuse	66
16.	Miscellaneous poisons	71

GENERAL TOXICOLOGY

1. Which one of the following is a mechanical antidote?
    A. Animal charcoal
    B. Copper sulphate
    C. Sodium chloride
    D. Tannic acid

2. BAL is useful in the treatment of poisoning from
    A. Endrin
    B. Mercury
    C. Opium
    D. Phosphorous

3. EDTA is useful in the treatment of
    A. Hypercalcaemia
    B. Hypocalcaemic tetany
    C. Lead poisoning
    D. Mercury poisoning

4. Alkalinisation of urine increases the excretion of
    A. Amphetamine
    B. Opium
    C. Phenobarbiotone
    D. Quinine

5. Hemodialysis is extremely useful in treating the poisoning due to
    A. Amphetamine
    B. Barbiturates
    C. Cocaine
    D. Opium

[**Answers:** 1 – A, 2 – B, 3 – C, 4 – C, 5 - B]

6. Which one of the following viscera is not commonly preserved for chemical analysis?
    A. Stomach and a part of the proximal part of the small intestine
    B. Liver (500 g) and half of each kidney
    C. Whole brain
    D. Blood

7. While dealing with a case of suspected poisoning, the doctor has the following medico legal duties to do except
    A. Intimation to the police
    B. Preservation of all the evidentiary materials for analysis
    C. To arrange for dying declaration if the patient is likely to die soon
    D. To issue the death certificate if demanded by the relatives

8. Which one of the following statement is not correct in relation to gastric lavage?
    A. The ingoing part of the tube is lubricated (True)
    B. The patient is to be right lateral position (False)
    C. Ensure that the poison is not corrosive mineral acid (True)
    D. Head is kept at a lower position than the rest of the body (True)

9. Which one of the following does not produce hypertension in overdose?
    A. Amphetamine
    B. Cocaine
    C. Nicotine
    D. Opiates

10. All of the following causes hypothermia in overdose except
    A. Barbiturates
    B. Carbon monoxide
    C. Opiates
    D. Salicylates

[**Answers:** 6 – C, 7 – D, 8 – B, 9 – D, 10 - D]

11. All of the following produce meiosis in overdose except
    A. Caffeine
    B. Cocaine
    C. Morphine
    D. Phenol

12. True about formalin is
    A. It can be used as preservative in alcohol poisoning.
    B. It is never used as preservative for chemical analysis.
    C. It is used as preservative in digitalis poisoning.
    D. None of the above

13. Narcotic drugs and psychotropic substance act was passed in the year of
    A. 1981
    B. 1983
    C. 1985
    D. 1986

14. Preservative used for toxicological specimen:
    A. 20% formalin
    B. Saturated sodium chloride
    C. 20% alcohol
    D. 10% alcohol

15. Gastric lavage is contraindicated in poisoning due to
    A. Sulphuric acid
    B. Organophosphorus compounds
    C. Arsenic
    D. Datura

16. Haemodialysis is useful in poisoning due to all of the following except
    A. Kerosine oil
    B. Barbiturates
    C. Alcohol
    D. Aspirin

[**Answers:** 11 – B, 12 – B, 13 – C, 14 – B, 15 – A, 16 - A]

17. Sodium fluoride may be useful for preservation of
    A. Cyanide
    B. Arsenic
    C. Alcohol
    D. Urine

18. Activated charcoal is used in
    A. Alcohol
    B. Barbiturates
    C. Heavy metal poisoning
    D. Lead poisoning

19. Gastric lavage is contraindicated in which of the following poisoning?
    A. Organophosphorus
    B. Datura
    C. Kerosene
    D. Copper sulphate

20. A sample to look for uric acid crystals (Gouty tophos) would be submitted to the pathology laboratory in
    A. Formalin
    B. Distilled water
    C. Alcohol
    D. Normal saline

21. Gastric lavage is indicated in all cases of acute poisoning because of
    A. Fear of aspiration
    B. Danger of cardiac arrest
    C. Danger of respiratory arrest
    D. Inadequate ventilation

22. A hypostasis of red-brown or deep blue in colour is suggestive of poisoning due to
    A. Nitrates
    B. Carbon monoxide
    C. Cyanides
    D. Barbiturates

[**Answers:** 17 – C, 18 – B, 19 – C, 20 – C, 21 – A, 22 - A]

23. The only disadvantage of activated charcoal administration in poisoning is that it is
    A. Expensive
    B. Unreliable in its effect
    C. Unpalatable
    D. Carcinogenic

24. One of the following viscera need not be preserved as a routine for chemical analysis is
    A. Heart
    B. Stomach
    C. Liver
    D. Kidney

25. Deep blue colour of hypostasis is seen in death due to poisoning by
    A. Potassium cyanide
    B. Phosphorus
    C. Aniline dye
    D. Carbon monoxide

26. Under the NDPS Act following drugs are included, expect
    A. Opium
    B. Hashish
    C. Amphetamine
    D. Alcohol

27. Delirium is seen in all of the following poisoning, except
    A. Datura
    B. Nux vomica
    C. Cocaine
    D. Lead

28. A dead body is having cadaveric lividity of bluish green color. The most likely cause of death is by poisoning due to
    A. Hydrocyanic acid
    B. Hydrogen sulphide
    C. Oleander
    D. Sodium nitrite

[Answers: 23 – C, 24 – A, 25 – C, 26 – D, 27 – B, 28 - B]

29. A dead body with suspected poisoning is having hypostasis of red brown in color. It is suggestive of poisoning due to
    A. Nitrite/ Aniline
    B. Carbon monoxide
    C. Cyanide
    D. Barbiturate

30. True about household emetics are all, except
    A. Ipecac syrup is potent and safe
    B. NaCl solution in warm water is the safest
    C. Apomorphine is effective orally
    D. Tickling the fauces with a tongue depressor is the best method

31. Activated charcoal is used in
    A. Alcohol poisoning
    B. Barbiturate poisoning
    C. Heavy metal poisoning
    D. Lead poisoning

32. Universal antidote consists of
    A. Activated charcoal
    B. Copper sulfate
    C. Egg white
    D. Starch

33. BAL is useful in treating poisoning due to all, except
    A. Lead
    B. Mercury
    C. Cadmium
    D. Arsenic

[Answers: 29 – A, 30 – C, 31 – B, 32 – A, 33 - C]

34. Drug containing two sulfhydryl groups in a molecule is
    A. BAL
    B. EDTA
    C. Penicillamine
    D. None

35. Urinary alkalization increases urine elimination of all the following drugs, except
    A. Salicylate
    B. Methotrexate
    C. Amphetamine
    D. Phenobarbitol

36. Alcohol, salicylates and pilocarpine can be used as
    A. Chelating agents
    B. Diaphoretics
    C. Purging
    D. Force alkaline dieresis

37. Hemodialysis is done in
    A. Organophosphorus poisoning
    B. Diazepam overdose
    C. Methanol poisoning
    D. CuSO$_4$ poisoning

38. Charcoal hemoperfusion is useful in which poisoning?
    A. Barbiturates
    B. Methanol
    C. Salicylate
    D. Digoxin

[**Answers:** 34 – A, 35 – C, 36 – B, 37 – C, 38 - A]

39. Hemodialysis is not used in case of poisoning with
    A. Salicylates
    B. Methanol
    C. Barbiturates
    D. Benzodiazepines

[**Answers:** 39 – D]

AGRICULTURAL POISONS

1. Organophosphorous poison is absorbed through
   A. Gastrointestinal tract
   B. Lungs
   C. Skin
   D. All of the above

2. Organophosphorous compounds inhibits
   A. Acetylcholine esterase
   B. Cytochrome oxidase
   C. Sulphydril enzymes
   D. None of the above

3. Muscarine- like effect of Acetyl choline include
   A. Ataxia
   B. Excess salivation
   C. Tremors
   D. Twitching of muscles

4. All of the following are the features of organophosphorus poisoning except
   A. Excessive salivation
   B. Loss of muscular co-ordination
   C. Low blood pressure
   D. Rhinorrhoea

5. The following are the correct examples of Organophosphorous compounds except
   A. DDT
   B. HETP
   C. OMPA
   D. TEPP

[**Answers:** 1 – D, 2 – A, 3 – B, 4 – B, 5 - A]

6. Which one of the following is known as 'Plant Penicillin'?
    A. DDT
    B. Endrin
    C. Parathion
    D. Malathion

7. Which one of the following is not an aryl phosphate?
    A. Parathion
    B. Malathion
    C. Follidol
    D. Tik-20

8. Organophosphorus insecticides are all, except
    A. Chlorpyriphos
    B. Gardona (tetrachlorvinphos)
    C. Dimethoate
    D. Diethyltoludamide (DEET)

9. All are features of organophosphorus poisoning, except
    A. Mydriasis
    B. Bradycardia
    C. Lacrimation
    D. Sweating

10. A 5 year old child presents with confusion, increased salivation, fasiculations, miosis, tachycardia and hypertension. Poison that can cause these manifestations is
    A. Opium
    B. Organophosphorus
    C. Datura
    D. Arsenic

[**Answers:** 6 – B, 7 – B, 8 – D, 9 – A, 10 - B]

11. Chromolacryorrhea is seen in poisoning with
    A. Cobra
    B. Organophosphorus
    C. Datura
    D. Carbolic acid

12. Estimation of plasma cholinesterase levels may be helpful in the management of poisoning with
    A. Datura
    B. Barbiturate
    C. Organophosphorus
    D. Opium

13. In a child with organophosphorus poisoning, following is the correct order of priority in management
    A. Pralidoxime, diazepam, atropine, clear airway
    B. Clear airway, atropine, diazepam, pralidoxime
    C. Diazepam, atropine, clear airway, pralidoxime
    D. Atropine, pralidoxime, diazepam, clear airway

14. A farmer visiting an orchard gets unconscious, excessive salivation, constricted pupils and fasciculation of muscles. Treatment is started with
    A. Atropine
    B. Neostigmine
    C. Physostigmine
    D. Adrenaline

15. Cholinesterase reactivator which goes to the brain is
    A. Pralidoxime
    B. Obidoxime
    C. Diacetyl monoxime
    D. None

[**Answers:** 11 – B, 12 – C, 13 – B, 14 – A, 15 - C]

16. Cholinesterase reactivators are ineffective in case of
    A. Baygon
    B. Parathion
    C. Malathion
    D. Tik 20

17. Which of the following is not a phase of organophosphorus poisoning?
    A. Acute cholinergic phase
    B. Intermediate syndrome
    C. Organophosphorus induced delayed polyneuropathy
    D. Late onset proximal myopathy

18. An 'intermediate syndrome' has been associated with
    A. Organophosphates
    B. Opium
    C. Cocaine
    D. Alphos

19. A patient is admitted with acute organophosphates insecticide poisoning, develops ptosis, inability to lift the head and difficulty in breathing on the third day. The most likely diagnosis is
    A. Hypokalemia
    B. Inflammatory polyneuropathy
    C. Intermediate syndrome
    D. Polymyositis

20. Polychlorinated hydrocarbon is
    A. Parathion
    B. Malathion
    C. Diazinon
    D. Endrin

[**Answers:** 16 – A, 17 – D, 18 – A, 19 – C, 20 - D]

21. Garlic odor around the nostrils and mouth is indicative of poisoning with
    A. Cyanide
    B. Organophosphorus
    C. Carbolic acid
    D. Aluminum phosphide

22. In aluminum phosphide poisoning, which is not true?
    A. Accumulation of acetylcholine at neuromuscular junction
    B. Cytochrome oxidase inhibition
    C. Phosphine formation
    D. Metabolic acidosis

23. A case of poisoning was brought to the casualty, gastric lavage turned black when it was heated after being treated with silver nitrate. The poisoning is most likely
    A. Tik 20
    B. Alphos
    C. Malathion
    D. Parathion

24. In treatment of Alphos poisoning, magnesium sulphate acts as
    A. Adjuvant
    B. Stabilizer
    C. Preservative
    D. Purgative

[**Answers:** 21 – D, 22 – A, 23 – B, 24 - B]

CORROSIVE POISONS

1. Which one of the following is the systemic action of Oxalic acid poisoning?
    A. Heaptic failure
    B. Hypoglycaemia
    C. Hypercalcaemia
    D. Renal failure

2. Which one of the following is the antidote for oxalic acid poisoning?
    A. Animal charcoal
    B. Aluminum oxide
    C. Calcium lactate
    D. Magnesium oxide

3. Postal envelop shaped crystals in the urine are the features of poisoning from
    A. Carbolic acid
    B. Oxalic acid
    C. Nitric acid
    D. Sulphuric acid

4. Hydroquinone and Pyrocatechol are the metabolic byproducts of
    A. Acetyl salicylic acid
    B. Carbolic acid
    C. Nitric acid
    D. Oxalic acid

5. Which one of the following is an example of mineral acid?
    A. Acetyl salicylic acid
    B. Carbolic acid
    C. Nitric acid
    D. Oxalic acid

[**Answers:** 1 – D, 2 – C, 3 – B, 4 – B, 5 - C]

6. In relation to oxalic acid which one of the following statement is false?
    A. Acts both locally and systemically
    B. Is used extensively as a stain remover
    C. Causes hypercalcaemia
    D. When swallowed blackens the mucosa of the mouth and the oesophagus

7. Xanthoproteic reaction is a feature of poisoning from
    A. Sulphuric acid
    B. Nitric acid
    C. Hydrochloric acid
    D. Carbolic acid

8. Vitriolage is
    A. Consuming alcohol mixed with acids
    B. Committing suicide by consuming acids
    C. Throwing of acid on a person
    D. Gastric lavage done with Ryle's tube

9. Oxalic acid poisoning results in
    A. Hypernatremia
    B. Hypocalcaemia
    C. Hyperkalaemia
    D. Hyper pigmentation of the skin

10. Greenish discolouration of urine on standing is a feature of poisoning from
    A. Acetyl salycilic acid
    B. Carbolic acid
    C. Methyl salycilic acid
    D. Oxalic acid

[**Answers:** 6 – C, 7 – B, 8 – C, 9 – B, 10 - B]

11. Which one of the following is the feature of Sulphuric acid poisoning?
    A. Dark tongue with yellowish teeth
    B. Blackish tongue with chalky white teeth
    C. Bluish line at the junction of the teeth and the gums
    D. Sequestration and necrosis of the jaw

12. Tetany can be caused by poisoning with
    A. Carbamates
    B. Curare
    C. Oxaltes
    D. Strychnine

13. Putrefaction is retarded by
    A. Carbolic acid
    B. Oxalic acid
    C. Organophosphorus poison
    D. Hydrochloric acid

14. Poison having local action only is
    A. Sulphuric acid
    B. Carbolic acid
    C. Oxalic acid
    D. Phosphorus

15. Antidote for mineral acid poisoning is
    A. $MgSO_4$
    B. $CuSO_4$
    C. $NaHCO_3$
    D. MgO

16. Common toxin through vegetables is
    A. Boric acid
    B. Carbolic acid
    C. Tartaric acid
    D. Oxalic acid

[**Answers:** 11 – B, 12 – C, 13 – A, 14 – A, 15 – D, 16 - D]

17. Ochronosis is seen in poisoning with
    A. HCl
    B. Boric acid
    C. Oxalic acid
    D. Carbolic acid

18. Leathery stomach is seen in poisoning with
    A. HCl
    B. H_2SO_4
    C. Carbolic acid
    D. Oxalic acid

19. 'Boiled lobster' appearance is seen in poisoning with
    A. Carbolic acid
    B. Boric acid
    C. Oxalic acid
    D. H_2SO_4 poisoning

20. Maximum damage to esophagus is with
    A. H_2SO_4
    B. Sodium hydroxide
    C. Acetic acid
    D. Nitric acid

[**Answers:** 17 – D, 18 – C, 19 – B, 20 - B]

INORGANIC METALLIC IRRITANT POISONS

1. Which one of the following mimics cholera in its manifestations?
   A. Arsenic poisoning
   B. Calotropis poisoning
   C. Dathura poisoning
   D. Lead poisoning

2. Aldrich – Mees line is a feature of chronic
   A. Arsenic poisoning
   B. Lead poisoning
   C. Phosphorous poisoning
   D. Thallium poisoning

3. Burtonian line is a feature of
   A. Arsenic poisoning
   B. Copper poisoning
   C. Mercury poisoning
   D. Lead poisoning

4. Porphyrinurea and chromodachryorrhoea are the features of
   A. Arsenic poisoning
   B. Copper poisoning
   C. Lead poisoning
   D. Zinc poisoning

5. Peripheral blood smear in plumbism is charecterized by
   A. Punctate basophylia
   B. Reticulocytosis
   C. Reduced platelets
   D. All of the above

[**Answers:** 1 – A, 2 – A, 3 – D, 4 – C, 5 - D]

6. Metal fume fever is a feature of poisoning from inhalation of fumes of
   A. Copper sub acetate
   B. Lead oxide
   C. Mercuric oxide
   D. Zinc oxide

7. Acute Iron poisoning can cause all of the following except
   A. Metabolic acidosis
   B. Constipation
   C. Convulsions
   D. Hepatic failure

8. Alopecia is a characteristic feature of poisoning with
   A. Arsenic
   B. Lead
   C. Mercury
   D. Thallium

9. Most common source of chronic arsenic poisoning in India is
   A. Well water
   B. Bottled water
   C. Sea fish
   D. Ayurvedic medicines

10. Long bones are required to be preserved in
    A. Arsenic poisoning
    B. Aconite poisoning
    C. Datura poisoning
    D. Smack poisoning

11. About 20 g of hair are required to be preserved in
    A. Arsenic poisoning
    B. Aconite poisoning
    C. Iodine poisoning
    D. Codeine poisoning

[Answers: 6 – D, 7 – B, 8 – D, 9 – A, 10 – A, 11 - A]

12. In burnt bones, the following can be detected:
    A. Arsenic
    B. Lead
    C. Organophosphorus compounds
    D. None of the above

13. Which poisoning retards putrefaction?
    A. Phosphorus
    B. Arsenic
    C. Mercury
    D. Lead

14. Acrodynia is seen in
    A. Mercury
    B. Lead
    C. Zinc
    D. Arsenic

15. All are the features of lead poisoning except
    A. Diarrhoea
    B. Abdominal pain
    C. Encephalopathy
    D. Nephropathy

16. Anaemia, punctate basophilia, constipation, blue line in the gum are characteristics of
    A. Opium addiction
    B. Arsenic poisoning
    C. Mercury poisoning
    D. Lead poisoning.

17. Mercury poison acts on
    A. Proximal convoluted tubules
    B. Distal convoluted tubules
    C. Loop of Henle
    D. Collecting ducts

[**Answers:** 12 – A, 13 – B, 14 – A, 15 – A, 16 – D, 17 - A]

18. The poison that can be detected in hair long after death is
    A. Lead
    B. Mercury
    C. Arsenic
    D. Cannabis

19. All of the following are features of chronic lead poisoning except:
    A. Encephalopathy
    B. Burtonian line
    C. Cutaneous blisters
    D. Constipation

20. In arsenic poisoning, greatest amount is found in
    A. Muscle
    B. Kidney
    C. Liver
    D. Brain

21. Reinsch test is used in diagnosis of poisoning due to
    A. Arsenic
    B. Lead
    C. Iron
    D. Copper sulfate

22. In death due to suspected poisoning where cadaveric rigidity is lasting longer than usual, it may be a case of poisoning due to
    A. Lead
    B. Arsenic
    C. Mercury
    D. Copper

23. 'Red velvety' stomach mucosa is seen in poisoning with
    A. Mercury
    B. Arsenic
    C. Lead
    D. Copper

[**Answers:** 18 – C, 19 – C, 20 – C, 21 – A, 22 – B, 23 - B]

24. Arsenic causes all, except
    A. Raindrop pigmentation
    B. Alopecia
    C. Palmar hyperkeratosis
    D. Blue line in gums

25. A middle aged man present with paraesthesia of hands and feet. Examination revealed presence of 'Mees' lines in the nails and raindrop pigmentation in the hands. The most likely diagnosis is
    A. Lead poisoning
    B. Arsenic poisoning
    C. Thallium poisoning
    D. Mercury poisoning

26. Raindrop pigmentation is seen in
    A. Arsenic poisoning
    B. Phosphorus poisoning
    C. Mercury poisoning
    D. Thallium poisoning

27. Fatty yellow liver is seen in poisoning with
    A. Arsenic
    B. Aconite
    C. Oxalic acid
    D. Mercury

28. In chronic arsenic poisoning, the following samples are useful for laboratory examination, except
    A. Nail clippings
    B. Hair samples
    C. Bone biopsy
    D. Blood sample

[**Answers:** 24 – D, 25 – B, 26 – A, 27 – A, 28 - D]

29. Following are true regarding chronic mercury poisoning, except
    A. Gingivostomatitis
    B. Brownish spot on anterior lens capsule
    C. Fine tremor is seen
    D. Inappropriate shyness and irritability

30. A factory worker presented with tremors, personality change and a blue line in gums. Probable diagnosis is chronic poisoning with
    A. Lead
    B. Mercury
    C. Arsenic
    D. Thallium

31. Hatter's shakes are seen in
    A. Lead poisoning
    B. Mercury poisoning
    C. Arsenic poisoning
    D. Copper poisoning

32. In mercury poisoning, brown reflex is from
    A. Anterior cornea
    B. Posterior cornea
    C. Anterior lens capsule
    D. Posterior lens capsule

33. Minamata Bay disease refers to chronic toxicity with
    A. Ergot
    B. Datura
    C. Organophosphorus
    D. Mercury

[**Answers:** 29 – C, 30 – B, 31 – B, 32 – C, 33 - D]

34. Plumbism is due to chronic poisoning with
    A. Arsenic
    B. Lead
    C. Mercury
    D. Copper

35. Commonest source of lead to cause increased blood level in children is from
    A. Air
    B. Lead paint dust
    C. Water
    D. Fruits

36. Lead inhibits which enzymes in the heme synthesis pathway?
    A. Aminolevulinate synthase
    B. Ferrochelatase and δ-ALA dehydratase
    C. Porphobilinogen deaminase
    D. Uroprophyrinogen decarboxylase

37. Basophilic stippling is seen in which of the following cells?
    A. Neutrophil
    B. RBC
    C. Basophils
    D. Eosinophils

38. Punctate basophilia is seen in poisoning with
    A. Lead
    B. Mercury
    C. Cadmium
    D. Potassium

[**Answers:** 34 – B, 35 – B, 36 – B, 37 – B, 38 - A]

39. In case of chronic lead poisoning, the levels of which of the following is elevated?
    A. Porphobilinogen
    B. δ-amino levulinic acid
    C. Bilirubin
    D. Urobilinogen

40. Copper sulphate poisoning manifests with
    A. Acute hemolysis
    B. High anion gap acidosis
    C. Peripheral neuropathy
    D. Rhabdomyolysis

41. Instead of penicillamine, following can be used in copper poisoning
    A. EDTA
    B. Desferrioxamine
    C. Succimer
    D. KMnO$_4$

42. A housewife ingests a rodenticide white powder accidentally. Her examination showed generalized flaccid paralysis and an irregular pulse. ECG shows multiple ventricular ectopics, generalized changes within ST-T. Serum potassium is 2.5 mEq/l. the most likely ingested poison is
    A. Barium carbonate
    B. Superwarfarins
    C. Zinc phosphide
    D. Aluminum phosphide

43. A person presents with acute poisoning, with chills and rigors similar to malaria. Most likely poisoning is with
    A. Mercury
    B. Zinc
    C. Red phosphorus
    D. Arsenic

[**Answers:** 39 – B, 40 – A, 41 – A, 42 – A, 43 - B]

44. Symptoms of metal fume fever resolve spontaneously within
    A. 6-12 hours
    B. 12-24 hours
    C. 24-36 hours
    D. 36-48 hours

45. The following drugs cause methemoglobinemia
    A. Aniline
    B. Dapsone
    C. Nitrates
    D. All

46. Patient with BP 90/60 mmHg, lips and peripheries are cyanosed, blood drawn was chocolate color. Diagnosis is
    A. Methemoglobinemia
    B. Hypovolemic shock
    C. Metal fume fever
    D. Alphos poisoning

[**Answers:** 44 – C, 45 – D, 46 - A]

INORGANIC NON-METALLIC IRRITANT POISONS

1. A poison which is luminescent and waxy and have a garlic smell is
    A. Alphos
    B. Ammonium bromide
    C. Opium
    D. Yellow phosphorus

2. CuSO4 was used as an antidote for
    A. Datura poisoning
    B. Cocaine poisoning
    C. Phosphorus poisoning
    D. Opium poisoning

3. A body is brought for autopsy with history of poisoning. On post-mortem, there is dark brown post-mortem staining and garlic odor in stomach. The poisoning is most likely due to
    A. Hydrocyanic acid
    B. Carbon dioxide
    C. Aniline dye
    D. Phosphorus

4. Yellow/fatty liver is characteristically seen in
    A. Datura poisoning
    B. Cocaine poisoning
    C. Phosphorus poisoning
    D. Opium poisoning

[Answers: 1 – D, 2 – C, 3 – D, 4 - C]

5. 'Phossy jaw' is seen in chronic poisoning with
    A. Datura
    B. Phosphorus
    C. Arsenic
    D. Thallium

6. Following are the properties of White phosphorus except
    A. Garlic smell
    B. Luminescent in dark
    C. Insoluble in water
    D. Amorphous

7. Side of the match box contain
    A. Red phosphorus and powdered glass
    B. Yellow phosphorus and diamond dust
    C. Lead carbonate and MgSO4
    D. Copper sulphate and charcoal

8. Tip of the match stick contain
    A. Potassium Chloride and antimony nitrite
    B. Potassium chlorate and antimony sulphide
    C. Sodium chloride and zinc sulphate
    D. Potassium cyanide and copper sulphate

9. Chemical antidote for phosphorus poisoning is
    A. $KMnO_4$
    B. Atropine
    C. Physostigmine
    D. $MgSO_4$

10. Fatal dose of white phosphorus is
    A. 1 - 2 mg
    B. 5 - 10 mg
    C. 15 - 30 mg
    D. 60 – 120 mg

[**Answers:** 5 – B, 6 – D, 7 – A, 8 – B, 9 – A, 10 - D]

ORGANIC PLANT IRRITANT POISONS

1. 'Suis' are prepared form
    A. Abrus precatorius
    B. Ricinus communis
    C. Nicotiana tobaccum
    D. Semicarpus anacardium

2. 'Marking Nut' is the common name of
    A. Abrus precatorius
    B. Argemone mexicana
    C. Semicarpus anacardium
    D. Strychnos Nux Vomica

3. In which one of the following, the toxic principle 'Amygdalin' is present?
    A. Apple seeds
    B. Castor seeds
    C. Dhatura seeds
    D. Marking nut

4. Ricin is obtained from
    A. Marking nut
    B. Poppy seed
    C. Castor seed
    D. Croton seed

5. A toxalbumin similar to viperine snake venom is present in the seed of
    A. Abrus precatorius
    B. Datura
    C. Ergot
    D. Croton tiglium

[**Answers:** 1 – A, 2 – C, 3 – A, 4 – C, 5 - A]

6. Toxic substance commonly used by washermen to put marks on clothes is
   A. Calotropis procera
   B. Plumbago rosea
   C. Semecarpus anacardium
   D. Croton tiglium

7. The case of the umbrella murder is related with
   A. Ricin
   B. Crotin
   C. Abrin
   D. Calotropin

8. Castor is also known as
   A. Ricinus communis
   B. Croton tiglium
   C. Abrus precatorius
   D. Calotropis

9. Fatal dose of semecarpus anacardium is
   A. 1 – 2 mg
   B. 1 – 2 g
   C. 5 – 10 mg
   D. 5 – 10 g

10. Seeds of capsicum annum are resemble in appearance to
    A. Castor seeds
    B. Croton seeds
    C. Datura seeds
    D. Abrus precatorius seeds

11. Hyderabadi goli is related with
    A. Datura
    B. Capsicum annum
    C. Abrus precatorius
    D. Semecarpus anacardium

[Answers: 6 – C, 7 – A, 8 – A, 9 – D, 10 – C, 11 - B]

12. St. Anthony's fire is related with
    A. Ergot
    B. Datura
    C. Castor
    D. Croton

13. Castor seeds are resemble in appearance to
    A. Datura seeds
    B. Capsicum seeds
    C. Abrus precatorius seeds
    D. Croton seeds

[**Answers:** 12 – A, 13 - D]

ORGANIC ANIMAL IRRITANT POISONS

1. Canthardine poisoning characteristically affects the
    A. Heart
    B. Kidney
    C. Liver
    D. Spinal cord

2. Priapism is a feature of
    A. Canthardine poisoning
    B. Bee sting
    C. Rattle snake poisoning
    D. Scorpion sting

3. The viper venom is predominantly
    A. Hepatotoxic
    B. Nephrotoxic
    C. Neurotoxic
    D. Vasculotoxic

4. The polyvalent antivenom available in India is not effective against
    A. Banded krait
    B. Common krait
    C. Common cobra
    D. Saw-scaled viper

5. Cantharides poisoning results from
    A. Biting
    B. Ingestion
    C. Injection
    D. Stinging

[**Answers:** 1 – B, 2 – A, 3 – D, 4 – A, 5 - B]

6. Which of the following snakes are poisonous?
   A. Krait
   B. Hydrophinae
   C. Cobra
   D. All

7. True of poisonous snakes are all, except
   A. Fangs present
   B. Belly scales are small
   C. Small head scales
   D. Grooved teeth

8. Cobra poison is
   A. Neurotoxic
   B. Mytoxic
   C. Carditoxic
   D. Vasculotoxic

9. Krait poison is
   A. Vasculotoxic
   B. Neurotoxic
   C. Cardiotoxic
   D. Hemotoxic

10. Venom of sea snake is mostly
    A. Neurotoxic
    B. Hemolytic
    C. Mytoxic
    D. Hepatotoxic

11. Cholinesterase is seen in venom of
    A. Elapids
    B. Vipers
    C. Sea snakes
    D. All

[**Answers:** 6 – D, 7 – B, 8 – A, 9 – B, 10 – C, 11 - A]

12. Lethal dose of krait venom is
    A. 3 mg
    B. 6 mg
    C. 12 mg
    D. 15 mg

13. Most characteristic feature of elapidae snake envenomation is
    A. Bleeding manifestation
    B. Neuro-paralytic symptoms
    C. Rhabdomyolysis
    D. Cardiotoxicity

14. A girl, otherwise healthy, sleeping on the floor suddenly develops nausea, vomiting, abdominal pain, quadriplegia at night. Diagnosis is
    A. Guillain Barre syndrome
    B. Poliomyelitis
    C. Krait bite
    D. Periodic paralysis

15. Muscle paralysis is caused by bite of
    A. Sea snake
    B. Krait
    C. Mamba
    D. Python

16. Following are recommended for primary management of a patient with snake bite, except
    A. Splinting and immobilization
    B. Keep the site of bite below heart
    C. Wash with soap and water
    D. Reassure the patient

[**Answers:** 12 – B, 13 – B, 14 – C, 15 – A, 16 - C]

17. Ligature pressure that should be used to resist spread of poison in elapidae poisoning is
    A. <10 mm Hg
    B. 20-30 mm Hg
    C. 50-70 mm Hg
    D. >100 mm Hg

18. Polyvalent snake vaccines contains immunoglobins against all, except
    A. Ophiophagus Hannah
    B. Naja naja
    C. Daboia rusellii
    D. Bungarus caeruleus

[**Answers:** 17 – C, 18 - A]

SOMNIFEROUS POISONS

1. All of the following are seen in Opium poisoning except
    A. Flaccidity of the muscles
    B. Dry skin
    C. Hypotension
    D. Pinpoint pupils

2. The characteristic feature of opium poisoning is
    A. Dryness of the skin and lack of sweating
    B. Hypertension
    C. Pin point pupils
    D. Tachypnoea

3. All of the following are the features of morphine poisoning except
    A. Bradycardia
    B. Exaggerated reflexes
    C. Pinpoint pupils
    D. Profound sweating

4. The following are the phenanthrene group of alkaloids present in opium except
    A. Codeine
    B. Morphine
    C. Papavarine
    D. Thebain

5. The term 'Brown Sugar' refers to impure form of
    A. Cannabis
    B. Cocaine
    C. Heroin
    D. Morphine

[**Answers:** 1 – B, 2 – C, 3 – B, 4 – C, 5 – C]

6. Which of the following is least narcotic?
    A. Morphine
    B. Codeine
    C. Thebane
    D. Papaverine

7. Which of these is not an opioid agonist?
    A. Heroin
    B. Ketamine
    C. Methadone
    D. Fentanyl

8. All are alkaloids, except
    A. Morphine
    B. Physostigmine
    C. Atropine
    D. Abrine

9. A 28 years old male patient is brought to casualty in comatose state with pin point pupils, reduced respiratory rate and bradycardia. Most likely diagnosis is
    A. Tricyclic antidepressant poisoning
    B. Opioid poisoning
    C. Benzodiazepine poisoning
    D. Organophosphorus poisoning

10. All are features of acute morphine poisoning, except
    A. Pin point pupil
    B. Hyperpyrexia
    C. Fall in blood pressure
    D. Slow labored breathing

[**Answers:** 6 – D, 7 – B, 8 – D, 9 – B, 10 - B]

11. Pin-point pupils are seen in
    A. Atropa belladona poisoning
    B. Opium poisoning
    C. Alphos poisoning
    D. Datura poisoning

12. Acute poisoning of narcotics present with
    A. Hypertension
    B. Hyperventilation
    C. Slow and shallow respiration
    D. Dilated pupils

13. Most common feature of opiate poisoning is
    A. Respiratory depression
    B. Hypotension
    C. Bradycardia
    D. Hypothermia

14. Opium poisoning is treated with
    A. Naloxone
    B. Atropine
    C. Neostigmine
    D. Physostigmine

15. Marquis test is done for
    A. Mercury poisoning
    B. Arsenic poisoning
    C. Morphine poisoning
    D. Cyanide poisoning

[**Answers:** 11 – B, 12 – C, 13 – A, 14 – A, 15 - C]

INEBRIANT POISONS

1. The best antidote for methyl alcohol poisoning is
    A. Alkaline diuresis
    B. Ethyl alcohol
    C. I.V fructose
    D. Mannitol IV

2. Antabuse
    A. Acts on the renal tubules to enhance ethanol excretion
    B. Acts on the addiction center of the brain
    C. Blocks oxidation of ethanol metabolic products
    D. Decreases the intestinal absorption of ethanol

3. Preservative used for the viscera in alcohol poisoning is
    A. Formalin
    B. Glycerin
    C. Sodium fluoride
    D. Saturated solution of Sodium chloride

4. Mc Ewan sign is seen in coma due to
    A. Alcohol
    B. Barbiturates
    C. Diabetes
    D. Opium

5. 'Mickey Finn' (Knockout drops) is a combination of Alcohol and
    A. Antihistamines
    B. Benzodiazapines
    C. Chloral hydrate
    D. Methaquolone

[**Answers:** 1 - B, 2 – C, 3 – D, 4 – A, 5 - C]

6. Mc Evan's sign in alcohol poisoning refers to changes in the
   A. Blood pressure
   B. Pupillary size
   C. Reflexes
   D. Respiratory rate

7. The duration of action of Phenobarbitone is
   A. Long
   B. Intermediate
   C. Short
   D. Ultra short

8. Drug automatism is seen with
   A. Alcohol
   B. Barbiturates
   C. Diazepam
   D. Morphine

9. Respiratory acidosis, blindness, colicky pain ending many times in death is associated with poisoning from
   A. Ethanol
   B. Methanol
   C. Cobra bite
   D. Scorpion sting

10. Sodium fluoride is used as preservative of blood in the estimation of
    A. Cyanide
    B. Mercury
    C. Arsenic
    D. Alcohol

11. Safe limit of alcohol consumption in males and females are
    A. 15 and 10 units/ week
    B. 18 and 15 units/ week
    C. 21 and 14 units/ week
    D. 25 and 18 units/ week

[Answers: 6 – B, 7 – A, 8 – B, 9 – B, 10 – D, 11 - C]

12. Black out is due to
    A. Alcohol intoxication
    B. Cocaine toxicity
    C. LSD toxicity
    D. Cyanide poisoning

13. In holiday heart syndrome, most common feature seen is
    A. Atrial fibrillation
    B. Atrial flutter
    C. Ventricular fibrillation
    D. Ventricular flutter

14. Arrhythmia most commonly associated with alcohol binge in the alcoholics
    A. Ventricular fibrillation
    B. Ventricular premature contraction
    C. Atrial flutter
    D. Atrial fibrillation

15. Korsakoff's psychosis is seen in
    A. CRF
    B. Chronic alcoholism
    C. Marasmus
    D. Cirrhosis

16. Disulfiram is useful in
    A. Alcohol dependence
    B. Heroin dependence
    C. Cocaine dependence
    D. Cannabis dependence

[**Answers:** 12 – A, 13 – A, 14 – D, 15 – B, 16 - A]

17. Disulfiram
    A. Inhibits alcohol dehydrogenase
    B. Inhibits aldehyde dehydrogenase
    C. Both A and B
    D. Inhibits phosphodiesterase

18. Drug that inhibits aldehyde dehydrogenase is
    A. Disulfiram
    B. Phenytoin
    C. Valproate
    D. Erythromycin

19. Disulfiram like reaction is caused by
    A. Acamprostate
    B. Metronidazole
    C. Tetracycline
    D. Digitalis

20. Most common symptom of alcohol withdrawal is
    A. Bodyache
    B. Tremor
    C. Diarrhea
    D. Rhinorrhea

21. CAGE questionnaire is used in
    A. Alcohol dependence
    B. Opiate poisoning
    C. Datura poisoning
    D. Barbiturate poisoning

22. Delirium tremens seen in
    A. Alcohol withdrawal
    B. Alcohol intoxication
    C. Opioid intoxication
    D. Opioid withdrawal

[**Answers:** 17 – B, 18 – A, 19 – B, 20 – B, 21 – A, 22 - A]

23. All are true about delirium tremens, except
    A. Normal sleep wake cycle
    B. Visual hallucinations
    C. Coarse tremors
    D. Clouding of consciousness

24. Delirium tremens is characterized by confusion associated with
    A. Autonomic hyperactivity and tremors
    B. Sixth nerve palsy
    C. Features of intoxication due to alcohol
    D. Korsakoff's psychosis

25. A 40 years old alcoholic is brought to the emergency with acute onset of seeing snakes all around him, not recognizing family members, violent behavior and tremulousness after having missed alcohol since 2 days. Examination reveals increased blood pressure, tremors, increased psychomotor activity, fearful effect, hallucinatory behavior, disorientation, impaired judgment and insight. He is most likely to be suffering from
    A. Alcoholic hallucinosis
    B. Delirium tremens
    C. Wernicke's encephalopathy
    D. Korsakoff's psychosis

26. The following is not a feature of Wernicke's encephalopathy
    A. Lateral rectus palsy
    B. Paralysis of conjugate gaze
    C. Pupillary dilatation
    D. Nystagmus

[**Answers:** 23 – A, 24 – A, 25 – B, 26 - C]

27. A 45 year male with a history of alcohol dependence presents with confusion, nystagmus and ataxia. Examination reveals 6th cranial nerve weakness. He is most likely to be suffering from
    A. Korsakoff's psychosis
    B. Wernicke's encephalopathy
    C. De Clearambault syndrome
    D. Delirium tremens

28. Wernicke-Korsakoff's syndrome is due to the deficiency of
    A. Pyridoxine
    B. Thiamine
    C. Vitamin B_{12}
    D. Riboflavin

29. Vitamin deficiency seen in alcoholic with dementia
    A. Thiamine
    B. Vitamin B12
    C. Riboflavin
    D. Pyridoxine

30. A 55 years old man presents with a 10 day history of confusion. His friend mentions that he drinks 15 units of alcohol per day. Which of the following strongly suggests a diagnosis of Korsakoff's psychosis?
    A. Delusional beliefs
    B. Poor long term memory
    C. Auditory hallucinations
    D. Confabulation

31. Area of the brain is usually not involved in Wernicke-Korsakoff syndrome is
    A. Periventricular gray matter
    B. Mammillary bodies
    C. Hippocampus
    D. Thalamus

[**Answers:** 27 – B, 28 – B, 29 – A, 30 – D, 31 - C]

32. True about alcohol paranoia is
    A. Tremors
    B. Fixed hallucinations
    C. Fixed delusions
    D. Wrist and foot drops

33. In India, driving under influence is considered at blood alcohol level of
    A. ≥20 mg%
    B. ≥30 mg%
    C. ≥50 mg%
    D. ≥100 mg%

34. Widmark's formula is used for measurement of blood levels of
    A. Benzodiazepines
    B. Barbiturates
    C. Alcohol
    D. Cocaine

35. The most reliable method of estimating blood alcohol level is
    A. Cavett's test
    B. Breath alcohol analyzer
    C. Gas liquid chromatography
    D. Thin layer chromatography

36. In methyl alcohol poisoning, CNS and cardiac depression and optic nerve atrophy are due to
    A. Formaldehyde and formic acid
    B. Acetaldehyde
    C. Pyridine
    D. Acetic acid

[**Answers:** 32 – C, 33 – B, 34 – C, 35 – C, 36 - A]

37. In contaminated liquor poisoning, all of the following are true, except
    A. Metabolic alkalosis
    B. Blindness
    C. Treatment is with ethanol
    D. Toxicity is due to methanol

38. True about blindness due to methanol poisoning are all, except
    A. Due to direct effect of formic acid
    B. Fomipezole, a specific alcohol dehydrogenase inhibitor is helpful
    C. Ethanol is useful to prevent blindness
    D. Gastric lavage is not helpful

39. All causes metabolic acidosis, except
    A. Methanol
    B. Ethanol
    C. Salicylate
    D. Isopropanol

40. High anion gap acidosis is seen in all the following, except
    A. Diabetic ketoacidosis
    B. Lactic acidosis
    C. Renal tubular acidosis
    D. Methanol poisoning

41. Antidote for ethylene glycol poisoning
    A. Methyl violet
    B. Fomepizole
    C. Fluconazole
    D. Ethyl alcohol

[**Answers:** 37 – A, 38 – D, 39 – D, 40 – C, 41 - B]

42. A 2 year old boy presents with fever for 3 days which responded to administration of paracetamol. Three days later he developed acute renal failure, marked acidosis and encephalopathy. His urine showed plenty of oxalate crystals. The blood anion gap and osmolal gap were increased. Most likely diagnosis is
    A. Paracetamol poisoning
    B. Ethylene glycol poisoning
    C. Severe malaria
    D. Hanta virus infection

[Answer: 42 – B]

DELIRIANT POISONS

1. Which one of the following preparation of cannabis is used as decoction?
     A. Bhang
     B. Charas
     C. Ganja
     D. Majoon

2. All of the following are the features of cocaine poisoning except
     A. Constricted pupils
     B. Hypothermia
     C. Marked sweating
     D. Confusion

3. The active principles, Atropine, Hyoscine and Hyoscyamine are seen in
     A. Abrus
     B. Dathura
     C. Cocaine
     D. Ricinus

4. Run Amok is associated with chronic poisoning from
     A. Cocaine
     B. Ricinus
     C. Semcarpus
     D. None of the above

5. 'Burnt rope' odour is associated with
     A. Tobacco
     B. Cannabis
     C. Carbon monoxide
     D. Camphor

[**Answers:** 1 – A, 2 – A, 3 – B, 4 – D, 5 – B]

6. The most potent form of Cannabis is
   A. Ganja
   B. Hashish
   C. Bhang
   D. Marijuana

7. The commonest route of intake of Cocaine by abusers is
   A. Inhalation
   B. Nasal insufflations
   C. Injection
   D. Ingestion

8. 'Indian hemp' refers to
   A. Cannabis sativa
   B. Erythroxylon coca
   C. Hyoscyamus niger
   D. Papaver somniferum

9. Following is not present in Datura
   A. Hyoscine
   B. Hyoscyamine
   C. Muscarine
   D. Atropine

10. The police brought a person from railway platform. He was talking irrelevant, had dry mouth with hot dry skin, dilated pupils, staggering gait and slurred speech. Most probable diagnosis is
    A. Alcoholic intoxication
    B. Datura poisoning
    C. Organophosphorus poisoning
    D. Aconite poisoning

[**Answers:** 6 – B, 7 – B, 8 – A, 9 – C, 10 - B]

11. All are true about atropine poisoning, except
    A. Dilated pupils
    B. Decreased temperature
    C. Dysarthria
    D. Dysphagia

12. A boy presented with bronchodilatation, increased temperature, constipation and tachycardia. Probable diagnosis is poisoning with
    A. Mushroom
    B. Atropine
    C. Penicillamine
    D. Organophosphorus

13. Following is used for treatment of belladonna poisoning
    A. Neostigmine
    B. Physostigmine
    C. Magnesium
    D. Atropine sulphate

14. Most common substance abuse in India
    A. Cannabis
    B. Tobacco
    C. Alcohol
    D. Opium

15. Active component of ganja is
    A. Tetrahydrocannabinol
    B. LSD
    C. N-methyl tryptophan
    D. Ricin

[**Answers:** 11 – B, 12 – B, 13 – B, 14 – A, 15 - A]

16. Nasal swabs are preserved in
    A. Drowning
    B. Anaphylaxis
    C. Cocaine poisoning
    D. Arsenic poisoning

17. Following are complications of cocaine poisoning, except
    A. Angina and myocardial infarction
    B. Epileptic seizures
    C. Hypothermia
    D. Hypertension

18. An addicted patient presenting with visual and tactile hallucinations, has black staining of tongue and teeth. The agent is
    A. Cocaine
    B. Cannabis
    C. Heroin
    D. Opium

19. A person feels that grains of sand are lying under the skin or some small insects are creeping on the skin giving rise to itching sensation, the condition is seen in
    A. Cocaine poisoning
    B. Organophosphorus poisoning
    C. Morphine poisoning
    D. Alcohol withdrawal

20. Magnan's syndrome is seen with
    A. Cocaine
    B. Organophosphorus
    C. Snake bite
    D. Alcohol

[**Answers:** 16 – C, 17 – C, 18 – A, 19 – A, 20 - A]

CARDIAC POISONS

1. Characteristic symptoms of aconite poisoning is
    A. Increased salivation
    B. Hypertension
    C. Tingling and numbness
    D. Hyperthermia

2. Which one of the following poison is known as Mitha Zaher?
    A. Aconite
    B. Nicotine
    C. Digitalis
    D. Oleander

3. Which one of the following is an ideal homicidal poison?
    A. Nicotine
    B. Digitalis
    C. Quinine
    D. Aconite

4. Root of aconite is resemble in appearance to
    A. Potato
    B. Horseradish root
    C. Brinjal
    D. Cucumber

5. Fatal dose of tobacco is
    A. 1 – 2 g
    B. 5 – 10 g
    C. 2 – 4 g
    D. 15 – 30 g

[**Answers:** 1 – C, 2 – A, 3 – D, 4 – B, 5 - D]

6. An average cigarette delivers how much nicotine?
    A. 1 – 3 mg
    B. 50 – 100 mg
    C. 250 – 500 mg
    D. 500 – 750 mg

7. Yellow oleander is known as
    A. Cerbera Thevetia
    B. Cinchona
    C. Nerium Odorum
    D. Digitalis purpurea

8. Which one of the following is similar to digitalis in action?
    A. Abrin
    B. Crotin
    C. Ricin
    D. Thevotoxin

[**Answers:** 6 – A, 7 – A, 8 - D]

SPINAL POISONS

1. Which one of the following poison mimics tetanus in its manifestations?
    A. Calpotropis procera
    B. Dathura alba
    C. Semicarpus anacardium
    D. Strychnos Nux Vomica

2. Post-mortem caloricity is seen in
    A. Strychnine poisoning
    B. Organophosphorus poisoning
    C. Datura poisoning
    D. Ergot poisoning

3. Inhibitory neurotransmitter in spinal cord is
    A. GABA
    B. Serine
    C. Glutamate
    D. Acetylcholine

4. Respiratory centre depression is caused by all, except
    A. Opium
    B. Strychnine
    C. Barbiturates
    D. Gelsemium

5. Which poison is known as dog button?
    A. Opium
    B. Abrus precatorius
    C. Strychnos nux vomica
    D. Ricinus communis

[**Answers:** 1 – D, 2 – A, 3 – A, 4 – B, 5 - C]

6. Following are active principles of strychnos nux vomica except
   A. Strychnine
   B. Loganin
   C. Brucine
   D. Ricin

7. Hyper extension of body in strychnine poisoning is known as
   A. Opisthotonus
   B. Emprosthotonus
   C. Pleurosthotonus
   D. Risus sardonicus

8. Greek philosopher, Socrates was died due to
   A. Strychnine poisoning
   B. Conium maculatum poisoning
   C. Curare poisoning
   D. Datura poisoning

9. Which of the following poison is known as Hemlock?
   A. Strychnos nux vomica
   B. Curare
   C. Conium maculatum
   D. Abrus precatorius

10. Which one of the following is peripheral nerve poison?
    A. Curare
    B. Datura
    C. Opium
    D. Calotropis

[**Answers:** 6 – D, 7 – A, 8 – B, 9 – C, 10 - A]

ASPHYXIANTS

1. In cyanide poisoning Amyl nitrate is usually administered by
    A. Intra venous injection
    B. Inhalation
    C. Oral route
    D. Subcutaneous injection

2. Bitter almond smell is perceived in the poisoning from
    A. Cyanide
    B. Phosphorus
    C. Opium
    D. Aconite

3. All of the following are inhibitors of cytochrome oxidase, except
    A. Carbon monoxide
    B. Amytal
    C. Cyanide
    D. Azide

4. Cyanide poisoning kit does not contain
    A. Sodium thiosulfate
    B. Sodium nitrite
    C. Sodium bicarbonate
    D. Amyl nitrite

5. Amyl nitrite is antidote for
    A. Aconite poisoning
    B. Cyanide poisoning
    C. Carbon monoxide poisoning
    D. Hydrogen sulphide poisoning

[**Answers:** 1 – B, 2 – A, 3 – B, 4 – C, 5 - B]

6. At autopsy, the cyanide poisoning case will show the following features, except
    A. Characteristic bitter lemon smell
    B. Congested organs
    C. Skin may be pinkish or cherry red in color
    D. Erosion and hemorrhages in esophagus and stomach

7. Carbon monoxide poisoning causes
    A. Anemic hypoxia
    B. Histotoxic hypoxia
    C. Anoxic hypoxia
    D. Stagnant hypoxia

8. In CO poisoning, immediate emergency treatment
    A. 5% CO_2 inhalation
    B. 10% CO_2 inhalation
    C. High flow O_2
    D. Nitroglycerine

9. Cherry red color in post-mortem staining is a feature of poisoning with
    A. Nitrites
    B. Aniline
    C. Phosphorus
    D. CO

10. Post-mortem finding in CO poisoning is
    A. Cherry red hypostasis
    B. Intense cyanosis
    C. Excessive salivation
    D. Pin-point pupil

[**Answers:** 6 – A, 7 – A, 8 – C, 9 – D, 10 - A]

11. Sewer gas is
    A. Phosgene
    B. H₂S
    C. CO₂
    D. CO

12. Death caused in suicide by household things in Japan is due to the production of
    A. Acidic solution
    B. H₂S
    C. HCN gas
    D. CO

13. Blistering war gas is
    A. Chlorine gas
    B. Mustard gas
    C. HCN gas
    D. Tabun

14. Nerve gas is
    A. Methyl isocyanate
    B. Phosgene
    C. Diphenylchloroarsine
    D. Sarin

15. Most important and potential agent that can be used in bioterrorism
    A. Plague
    B. Small pox
    C. Tuberculosis
    D. Clostridium botulinum

[**Answers:** 11 – B, 12 – B, 13 – B, 14 – D, 15 - B]

FOOD POISONS

1. Which one of the following is not a manifestation of 'Botulism'?
    A. Diarrhoea
    B. Diplopia
    C. Dysphagia
    D. Descending paralysis

2. A 22 years old male had an outing with his friends and developed fever of 38.5 °C, diarrhea and vomiting after eating chicken salad 24 hours back. Two of his friends developed the same symptoms. The diagnosis is
    A. Salmonella enteritis poisoning
    B. Bacillus cereus poisoning
    C. Staphylococcus aureus poisoning
    D. Vibrio cholera poisoning

3. Which of the following is an exotoxin?
    A. E. coli toxin
    B. Proteus
    C. Pseudomonas
    D. Tetanus toxin

4. Mechanism of action of botulinum toxin is
    A. Synthesis of acetylcholine inhibited
    B. Reuptake of ACH is increased
    C. Blocks nicotinic receptors in muscle
    D. Blocks muscarinic receptors in brain

5. Botulinum affects all, except
    A. Neuromuscular junction
    B. Preganglionic junction
    C. Post ganglionic nerve
    D. CNS

[Answers: 1 – A, 2 – A, 3 – D, 4 – A, 5 - D]

6. Dysphagia, diplopia, Dysarthria are characteristic symptoms of food poisoning due to
   A. Staphylococcus aureus
   B. Clostridium botulinum
   C. Salmonella typhimurium
   D. Bacillus cereus

7. An 18 years old male presented with acute onset descending paralysis and blurring of vision of 3 days duration. On examination, the patient has quadriparesis with areflexia. Both the pupils are non-reactive. The most probable diagnosis is
   A. Poliomyelitis
   B. Botulism
   C. Diphtheria
   D. Porphyria

8. Which poisoning can be prevented by an antitoxin?
   A. Staphylococcus aureus
   B. Clostridium botulinum
   C. Salmonella typhimurium
   D. Bacillus cereus

9. Botulinum toxin is used for the treatment of
   A. Blepharospasm
   B. Risus sardonicus
   C. Strabismus
   D. All

10. In Kesari dhal poisoning due to Lathyrus sativus, the active principle is
    A. Pyrrozolidine
    B. BOAA
    C. Argemone oil
    D. Pilocarpine

[**Answers:** 6 – B, 7 – B, 8 – B, 9 – D, 10 - B]

11. Lathyrism is seen with eating of
    A. Red gram
    B. Kesari dhal
    C. Mushrooms
    D. Sausages

12. The drug of choice for mushroom poisoning is
    A. Atropine
    B. Physostigmine
    C. Adrenaline
    D. Carbachol

13. Epidemic dropsy is caused by
    A. Kesari dhal
    B. Argemone oil
    C. Poisonous mushrooms
    D. Shell fish

14. Toxin responsible for epidemic dropsy is
    A. BOAA
    B. Aflatoxin
    C. Sanguinarine
    D. Pyrrozolidine

[**Answers:** 11 – B, 12 – A, 13 – B, 14 - C]

DRUG DEPENDENCE AND ABUSE

1. Which of the following can help in the diagnosis of 'Body packer's syndrome'?
    A. Abdominal X-ray
    B. C.T Scan
    C. Ultrasound
    D. All of the above

2. The term 'Liquid gold' refers to
    A. Molten Gold
    B. Amalgam of Mercury and Gold
    C. Urine of Amphetamine abuser
    D. Urine of jaundiced patient

3. Treatment of acute alcohol withdrawal is
    A. Diazepam
    B. Bupropion
    C. Disulfiram
    D. Acamprosate

4. Drug not used in treatment of alcohol dependence is
    A. Diazepam
    B. Disulfiram
    C. Acamprosate
    D. Naltrexone

5. Alcohol anti-craving agents are all, except
    A. Lorazepam
    B. Clonidine
    C. Acamprosate
    D. Naltrexone

[Answers: 1 – D, 2 – C, 3 – A, 4 – A, 5- A]

6. Yawning is a common feature of
    A. Alcohol withdrawal
    B. Cocaine withdrawal
    C. Cannabis withdrawal
    D. Opioid withdrawal

7. All are true of opioid withdrawal, except
    A. Yawning
    B. Hallucinations
    C. Lacrimation
    D. Piloerection

8. Boy is having diarrhea, rhinorrhea and sweating, most probable diagnosis is
    A. Cocaine withdrawal
    B. Heroin withdrawal
    C. Marijuana withdrawal
    D. LSD withdrawal

9. In patients with substance abuse, drug not used is
    A. Naltrexone
    B. Naloxone
    C. Clonidine
    D. Chlorpromazine

10. Not used for treatment of heroin detoxification
    A. Disulfiram
    B. Buprenorphine
    C. Clonidine
    D. Lofexidine

[**Answers:** 6 – D, 7 – B, 8 – B, 9 – D, 10 - A]

11. The drug which is used for long term maintenance in opioid addiction
    A. Nalorphine
    B. Naloxone
    C. Butarphanol
    D. Methadone

12. Alternative to methadone for maintenance treatment of opiate dependence
    A. Diazepam
    B. Chlordiazepoxide
    C. Buprenorphine
    D. Dextropropoxyphene

13. Tolerance is seen in all, except
    A. Morphine
    B. Amphetamine
    C. Cocaine
    D. Barbiturates

14. Amotivational syndrome is seen with
    A. Heroin
    B. Cannabis
    C. Cocaine
    D. Clonidine

15. Symptomatic treatment is only required in withdrawal syndrome caused by
    A. Cannabis
    B. Morphine
    C. Alcohol
    D. Cocaine

[**Answers:** 11 – D, 12 – C, 13 – C, 14 – B, 15 - A]

16. Cocaine abuse is much similar to which abuse?
    A. Cannabis
    B. Nicotine
    C. Heroin
    D. Amphetamine

17. Paranoid schizophrenia is mimicked by intake of
    A. Amphetamine
    B. Heroin
    C. Cannabis
    D. Alcohol

18. True about MDMA is
    A. Ecstasy is another name for it.
    B. It is a cocaine congener
    C. Causes parkinsonism like symptoms
    D. Methadone is used to treat withdrawal symptoms.

19. A young city dweller presented with history of drug abuse and complaining of change in perception, like hearing sights and seeing sounds. Substance responsible for this is
    A. LSD
    B. Phencyclidine
    C. Cocaine
    D. Amphetamine

20. Rave drug is
    A. Cannabis
    B. Cocaine
    C. Heroin
    D. Methamphetamine

[**Answers:** 16 – D, 17 – A, 18 – A, 19 – A, 20 - D]

21. Drug used in the prophylaxis of nicotine addiction is
    A. Diazepam
    B. Naloxone
    C. Bupropion
    D. Acamprosate

22. All are used in nicotine de-addiction, except
    A. Bupropion
    B. Clonidine
    C. Nicotine gum
    D. Buspirone

[**Answers:** 21 – C, 22 - D]

MISCELLANEOUS POISONS

1. Aspirin is fatal at the blood levels of
   A. 5 mg%
   B. 20 mg%
   C. 50 mg%
   D. 100 mg%

2. A female age 26 years accidentally takes 100 tablets of paracetamol. Treatment of choice is
   A. Lavage with charcoal
   B. Dialysis
   C. Alkaline dieresis
   D. Acetylcysteine

3. N-acetyl-cysteine is antidote for toxicity with
   A. Benzodiazepine
   B. Barbiturates
   C. Acetaminophen
   D. Amphetamine

4. Treatment of acute iron toxicity is
   A. EDTA
   B. BAL
   C. Desferrioxamine
   D. Penicillamine

5. Antidote for benzodiazepine poisoning is
   A. Naloxone
   B. Atropine
   C. Flumazenil
   D. N-acetyl-cysteine

[Answers: 1 – D, 2 – D, 3 – C, 4 – C, 5 - C]

6. A patient ingested some unknown substance and presented with myoclonic jerks, seizures, tachycardia and hypotension. The ECG showed a heart rate of 120/ min with QRS interval of 0.16 seconds. The arterial blood revealed a pH of 7.25, PCO_2 of 30 mm Hg and HCO_3 of 15 mmol/l. Most likely cause of poisoning is
    A. Amanita phallodies
    B. Ethylene glycol
    C. Imipramine
    D. Phencyclidine

7. A woman consumes several tablets of amitriptyline. All of the following can be done, except
    A. Sodium bicarbonate infusion
    B. Gastric lavage
    C. Diazepam for seizure control
    D. Atropine as antidote

8. Dry wine is
    A. Methylated spirit
    B. Ethyl alcohol
    C. Chloral hydrate
    D. Isopropyl alcohol

9. Management of kerosene oil poisoning includes all, except
    A. Gastric lavage is done
    B. Bronchodilators are given
    C. Oxygen is given
    D. Corticosteroids are not beneficial

[**Answers:** 6 – C, 7 – D, 8 – C, 9 - A]

www.ingramcontent.com/pod-product-compliance
Lightning Source LLC
Chambersburg PA
CBHW080821170526
45158CB00009B/2487